Numbers

Peter Patilla

Heinemann Library
Des Plaines, Illinois

© 2000 Reed Educational & Professional Publishing
Published by Heinemann Library,
an imprint of Reed Educational & Professional Publishing,
1350 East Touhy Avenue, Suite 240 West
Des Plaines, IL 60018

Customer Service 1-888-454-2279

Designed by AMR
Illustrations by Art Construction and Jessica Stockham (Beehive Illustration)
Originated by HBM Print Ltd, Singapore
Printed and bound by South China Printing Co., Hong Kong/China

04 03 02 01 00
10 9 8 7 6 5 4 3 2 1

Library of Congress Cataloging-in-Publication Data
Patilla, Peter.
    Numbers / Peter Patilla.
        p. cm. — (Math links)
    Includes bibliographical references and index.
    Summary: Introduces basic number concepts, including numerals,
digits, number order, addition, subtraction, division,
multiplication, odd numbers, and even numbers.
    ISBN 1-57572-966-0 (lib. bdg.)
    1. Mathematics Juvenile literature. [1. Mathematics.]
    I.Title. II. Series: Patilla, Peter. Math links.
    QA40.5.P38 1999
    510—dc21   99-20369
                                                    CIP

Acknowledgments
The Publishers would like to thank the following for permission to reproduce photographs:
Allsport/Craig Prentis, p. 11; Trevor Clifford, p. 5, 7, 9, 10, 12, 13, 14, 15, 17, 18, 19, 20, 23, 24, 27, 28; Tony Stone Images/Paul Chesley, p. 29 top, Tony Stone Images /Jon Riley, p. 29 bottom.

Cover photo: Trevor Clifford

Our thanks to David Kirkby for his comments in the preparation of this book.

Every effort has been made to contact copyright holders of any material reproduced in this book. Any omissions will be rectified in subsequent printings if notice is given to the Publisher.

Some words in this book are in bold, **like this**. You can find out what they mean by looking in the glossary. Look for the answers to questions in the green boxes on page 32.

Contents

Numerals 4

Digits...6

Ordering Numbers 8

First to Last 10

Sets..12

Comparing Sets.................................14

Addition 16

Subtraction..18

Multiplication 20

Sharing 22

Odd and Even 24

Fractions 26

Large Numbers...................................28

Glossary 30

More Books to Read 31

Answers 32

Index 32

Numerals

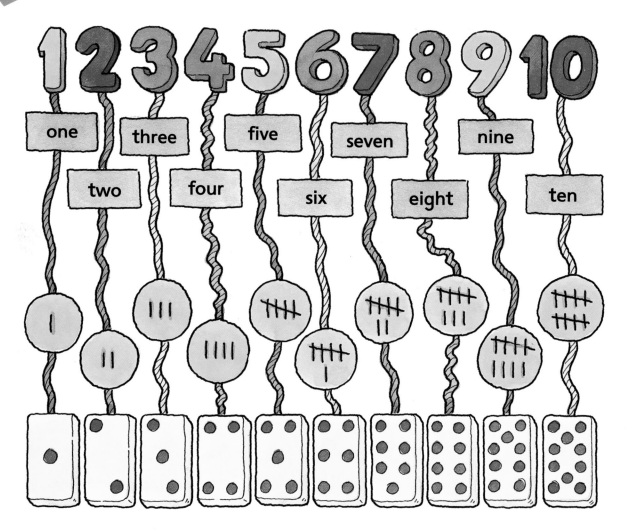

Numbers can be shown in lots of different ways.
We can use words, pictures, signs, or **symbols**.
They are all called **numerals**.

4

All kinds of numerals can be used to show numbers. Here are some you might see.

Look around you and find some numerals. Where did you find them?

5

Digits

0 zero
1 one
2 two
3 three
4 four
5 five
6 six
7 seven
8 eight
9 nine

These are 2 digit numbers.

These are 3 digit numbers.

18 42 136 207

There are 10 **digits**. They are the numbers 0, 1, 2, 3, 4, 5, 6, 7, 8, and 9. Digits are important because they are used to write other numbers.

6

Big numbers, little numbers, measurements, and money use digits. Some clocks use digits to tell the time. Calculators uses digits to give the answer.

Which digits do you see in the picture?

Ordering Numbers

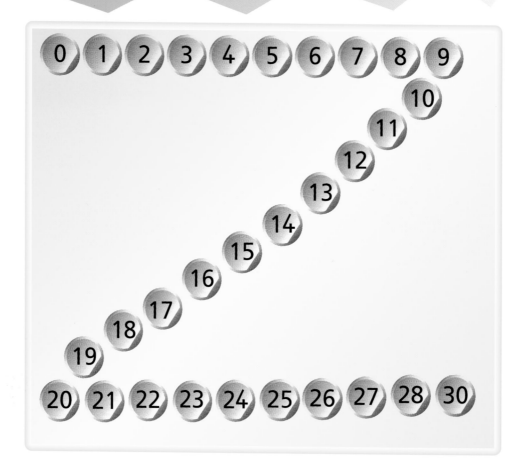

Numbers have an order. You can start anywhere and go forward or backward. Numbers go on and on. They never stop. It is important to know the order of numbers.

8

Numbers are in order in lots of places.

The numbers 1 to 12 are in order on some clocks.

The numbers 1 to 9 are in order on a telephone.

Where can you see numbers written in order
in the picture?

First to Last

first   middle   last

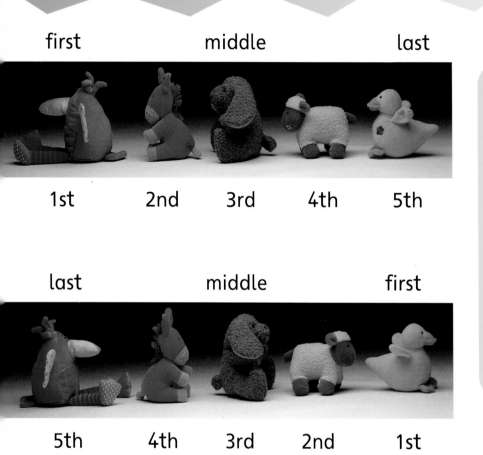

1st   2nd   3rd   4th   5th

last   middle   first

5th   4th   3rd   2nd   1st

1st	first
2nd	second
3rd	third
4th	fourth
5th	fifth
6th	sixth
7th	seventh
8th	eighth
9th	ninth
10th	tenth

Words such as first, second, and third tell the order of things in a line. Five things are ordered from 1st to 5th. Ten things are ordered to 10th.

10

Order is important in a race. When runners are close together, it can be hard to tell who is first.

Look at the picture. Who is first? Who is second? Who is last?

Sets

Things are often packed in groups of twos, threes, sixes, and twelves. A group of like things is called a set. Words such as **pair, half dozen**, and **dozen** tell us how many are in a set.

Things are sometimes put into patterns. The patterns help us count. The patterns also help show when some of the set is missing.

Look at the picture. How many are missing?

Comparing Sets

more   fewer   equal

more   less   equal

$8 > 6$ means 8 is **greater than** 6

$6 < 8$ means 6 is **less than** 8

$1 + 1 = 2$ means 1 and 1 equals 2

To compare two amounts or numbers, we can use words such as more, fewer, less, **equal**, and same. We can also use the signs **<**, **>**, and **=**.

To compare amounts, we can use the words
most, meaning more, and *least,* meaning fewer.

Which plate has the most sandwiches?
Which glass has the least juice?

Addition

$$4 + 5 = 9$$

$$3 + 4 + 2 = 9$$

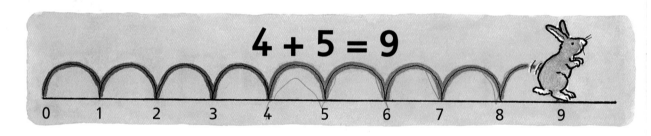

$$4 + 5 = 9$$

0 1 2 3 4 5 6 7 8 9

Addition is putting two or more parts together.
It is also jumping forward along a number line.
How many in all is called the **sum** or **total**.

In this game, the numbers on each corner add up to the number in the middle of the triangle.

Which numbers do you think are hidden by the children's fingers?

Subtraction

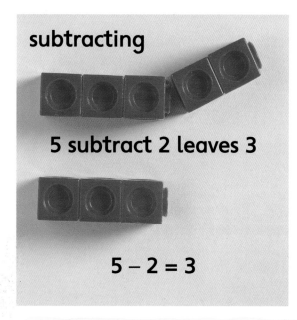

subtracting

5 subtract 2 leaves 3

5 − 2 = 3

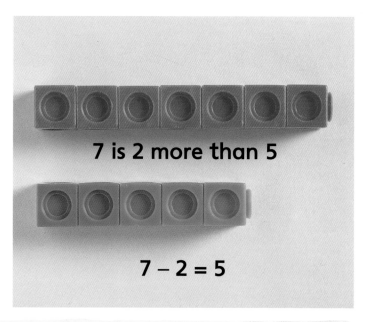

7 is 2 more than 5

7 − 2 = 5

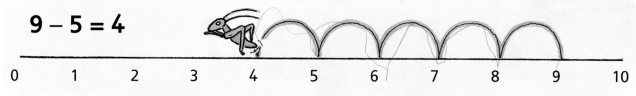

9 − 5 = 4

0   1   2   3   4   5   6   7   8   9   10

Subtraction (−) is taking away one number from another. It is also comparing two numbers to find the **difference** between them. Subtraction is also jumping backward along a number line.

18

Each child has a total of 10 beads. Each child is covering some of the beads.

How many beads does each child have on the table? How many is each child covering?

Multiplication

4 bundles
3 pens in each bundle

3 + 3 + 3 + 3

4 x 3 = 12

8 x 3 = 24

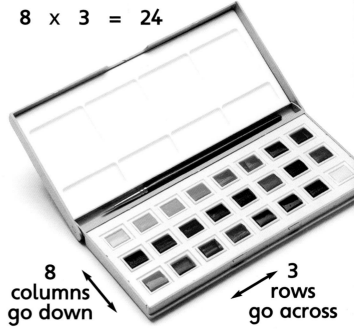

8
columns
go down

3
rows
go across

Multiplication (x) is adding together **equal** groups. It is adding numbers lots of times. Multiplication is a fast way to find how many in all.

When things are in rows and columns, they are arranged in equal groups. Equal groups of twos, threes, fours, or fives can easily be counted.

How many cookies are on the tray?

Sharing

9 cherries

3 equal groups, 9 ÷ 3 = 3

2 equal groups with one left over, 9 ÷ 4 = 2, remainder 1

Division (÷) is sharing something into **equal** groups. If one group is not equal to the others, then it is called a **remainder**.

These things can be divided into small groups of twos and threes.

Which of these things can be divided into equal groups of twos? Which can be divided into equal groups of threes?

Odd and Even

Even numbers

Odd numbers

even numbers: 2, 4, 6, 8, 10, 12, 14, 16
odd numbers: 1, 3, 5, 7, 9, 11, 13, 15

Even numbers can be put into twos exactly.
Odd numbers always have one left over when
put into twos. **Zero** by itself is not even or odd.
It means none.

24

Look at the last digit of large numbers.

If it is 0, 2, 4, 6, 8, then the number is even.

If it is 1, 3, 5, 7, 9, then the number is odd.

Look at the numbers in the picture. Which numbers are odd? Which numbers are even?

Fractions

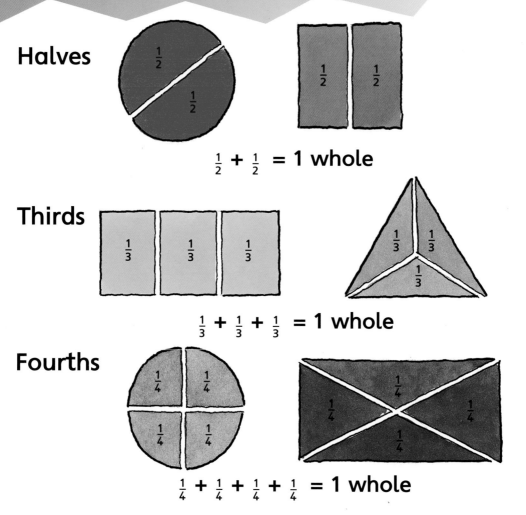

Halves

$\frac{1}{2} + \frac{1}{2} = 1$ whole

Thirds

$\frac{1}{3} + \frac{1}{3} + \frac{1}{3} = 1$ whole

Fourths

$\frac{1}{4} + \frac{1}{4} + \frac{1}{4} + \frac{1}{4} = 1$ whole

Whole shapes and amounts can be divided into **fractions**. Fractions are parts of a whole. Each part of the whole must be a fair share.

26

When you cut something up fairly, each fraction is the same size.

Is the whole pizza divided into equal fourths? Which fraction is missing from pizza 1? Which fraction is missing from pizza 2?

Large Numbers

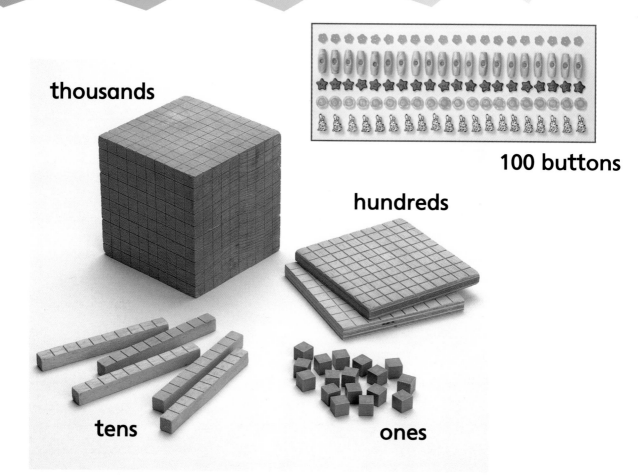

thousands

hundreds

100 buttons

tens

ones

Words such as **hundreds, thousands,** tens of thousands, and **millions** tell about some large numbers. Since numbers can go on and on, it can be hard to imagine very large numbers.

There are large numbers of people in schools, at sporting events, and on busy city streets.

Which picture shows thousands of people?

Glossary

addition (+) putting groups together to count how many there are in all

difference how many more one number is than another (subtraction)

digit any of the numbers 0, 1, 2, 3, 4, 5, 6, 7, 8, and 9

division (÷) finding how many equal groups are in a number

dozen 12 of anything

equal (=) the same in amount, size, or number

fraction part of a whole. Halves, thirds, and fourths are fractions.

greater than (>) when there is a larger amount of something

half dozen 6 of anything

hundred (100) ten tens, or a hundred ones

less than (<) when there is not as much of something

million (1,000,000) a thousand thousands

multiplication (×) adding the same number lots of times

numeral the way we write and show numbers

pair things grouped in twos

remainder what is left over after something is divided into equal groups

subtraction (−) finding the difference between two numbers

sum the answer when numbers are added together (addition)

symbol sign that stands for a word or action: add +; subtract −; multiply ×; divide ÷

thousand (1,000) ten hundreds or a hundred tens

total the whole amount; sum

zero (0) a digit. It means none.

More Books to Read

Kirkby, David. *Numbers*. Crystal Lake, Ill.: Rigby Interactive Library, 1996.

— *Number Play*. Crystal Lake: Rigby Interactive Library, 1996. An older reader can help you with this book.

Leedy, Loreen. *Fraction Action*. New York: Holiday House, 1994.

Murphy, Stuart J. *Divide and Ride*. New York: Harper Collins, 1997.

Patten, John M., Jr. *Numbers & Counting*. Vero Beach, Fla.: Rourke Corporation, 1996.

Schmandt-Besserat, Denise. *The History of Counting*. New York: William Morrow, 1997. An older reader can help you with this book.

Tabor, Nancy M. Grande. *Fifty on the Zebra (Cincuenta en la Cebra): Counting with Animals (Contando Con los Animales)*. Watertown, Mass.: Charlesbridge Publishing, 1994.

Answers

page 7	0, 1, 2, 3, 4, 5, 6, 7, 8, 9
page 9	telephone, clock, calculator, calendar
page 11	1st–number 275, 2nd–number 679, last–number 414
page 13	4 missing
page 15	plate on right; glass in front
page 17	girl–5, boy–8
page 19	boy has 4 on the table; he is covering 6 girl has 6 on the table; she is covering 4
page 21	12, (3 x 4), 4 groups of 3
page 23	twos–notebooks, pencils; threes–rubberbands, pencils; pencils can be grouped into twos or threes
page 25	odd: 49, 57, 31, 95, 15, 163 even: 62, 42, 12, 40
page 27	no pizza 1: missing one half pizza 2: missing one fourth
page 29	top picture

Index

columns 20, 21
digits 6–7, 30
greater than 14, 30
less than 14, 30
number line 16, 18
numerals 4–7, 30
ordering 8–11
pairs 12, 30
patterns 13
remainder 22–23, 30
rows 20–21